W9-DDX-669   2015

Nora the Naturalist's Animals

Tide Pool Animals

A⁺
Smart Apple Media

Published by Smart Apple Media, an imprint of Black Rabbit Books
P.O. Box 3263, Mankato, Minnesota 56002
www.blackrabbitbooks.com

Produced by David West ♟ Children's Books
7 Princeton Court, 55 Felsham Road, London SW15 1AZ

Designed and illustrated by David West

Copyright © 2013 David West Children's Books

Library of Congress Cataloging-in-Publication Data

West, David, 1956-
Tide pool animals / David West.
        p. cm. – (Nora the Naturalist's animals)
Includes index.
 ISBN 978-1-62588-004-8 (library binding)
 ISBN 978-1-62588-054-3 (paperback)
1. Tide pool animals–Juvenile literature. 2. Tide pool ecology–Juvenile literature. I. Title.
QL122.2.W464 2014
591.769'9–dc23
                                                                                2013009584

Printed in China
CPSIA compliance information: DWCB13CP
010313

9 8 7 6 5 4 3 2 1

Nora the Naturalist says:
I will tell you something
more about the animal.

Learn what this
animal eats.

Where in the
world is the
animal found?

Its size is revealed!

What animal group
is it – mammal, bird,
reptile, amphibian,
insect, or something
else?

Interesting facts.

Contents

Crabs 4

Lobsters 6

Shrimp 8

Crayfish 10

Fish 12

Starfish 14

Jellyfish 16

Octopus 18

Sea Urchins 20

Sea Shells 22

Glossary and Index 24

Nora the Naturalist says:
Crabs have an outer layer of armored shell, called an exoskeleton. Many crabs walk sideways.

Crabs

Sometimes you may find a crab hidden amongst the seaweed in tide pools. They are known for their powerful pincer claws.

Edible crab

Crabs feed on any food available, such as **mollusks**, worms, other crustaceans, and decaying plant matter.

Crabs can be found all over the world, in the oceans and on seashores.

Adult crabs may have a body width of up to 10 inches (25 cm) and weigh as much as 6.6 pounds (3 kg).

Crabs are members of the **crustacean** family.

Many types of crab, like this brown crab (also known as the edible crab) are delicious to eat.

Lobsters

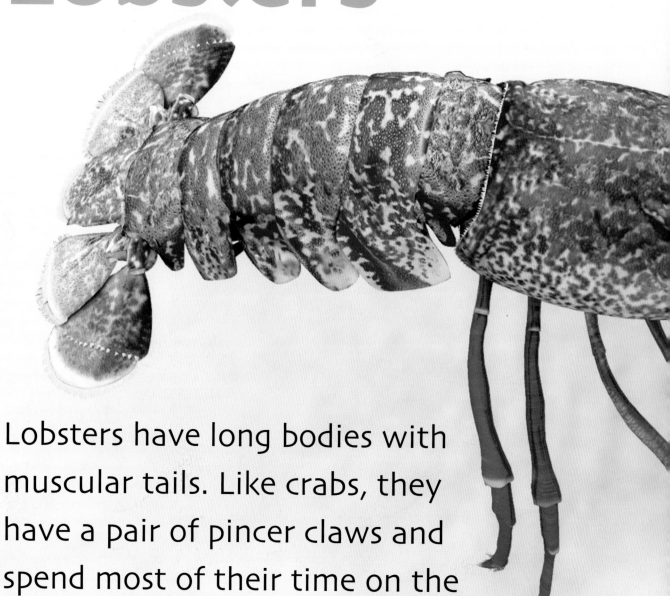

Lobsters have long bodies with muscular tails. Like crabs, they have a pair of pincer claws and spend most of their time on the sea floor, in crevices and burrows.

Lobsters typically eat live prey, such as fish, mollusks, other crustaceans, worms, and some plant life.

Lobsters are found in all the oceans.

In general, lobsters are 10–20 inches (25–50 cm) long.

Lobsters are members of the crustacean family.

Lobsters move by slowly walking on the sea floor. To escape a predator they swim backward quickly by curling and uncurling their tail at speeds of up to 11 mph (5 meters per second).

Nora the Naturalist says: Like crabs, lobsters must molt in order to grow. This means they shed their outer shell and wait a few hours until the new one hardens.

Blue lobster

Small shrimp will eat **plankton**. As they grow larger they scavenge on dead sea animals that have fallen to the sea floor.

 Shrimp are found in all the oceans of the world.

 Shrimp are about 0.79 inches (2 cm) long, but some shrimp (often called prawns) grow to over 9.8 inches (25 cm).

 Shrimp are members of the crustacean family.

 The tails of shrimp can be delicious to eat. They are caught around the world, and are also farmed for us to eat.

Shrimp

Shrimp

Shrimp are one of the most common sea animals to be found in tide pools. These tiny crustaceans can be quite difficult to see as they often have transparent bodies.

Nora the Naturalist says:
Unlike crabs and lobsters, shrimp have swimming legs and slender walking legs. They are more adapted for swimming than walking.

Nora the Naturalist says:
These spiny lobsters are not true lobsters. They don't have the large claws that true lobsters do.

They eat snails, clams, crabs, sea urchins, and dead sea animals.

Spiny lobsters are usually found in warm seas, including the Caribbean and the Mediterranean, and in Australasia and South Africa.

Crayfish can grow to be similar sizes as lobsters, 10–20 inches (25–50 cm) long.

Crayfish are members of the crustacean family.

It was recently discovered that spiny lobsters can navigate by using the Earth's magnetic field.

Crayfish

Although sometimes called spiny lobsters, or crawfish, these crustaceans are different from lobsters. They may be found near the shore, hiding under rocks.

Fish

All sorts of fish can be found in tide pools. Some are juveniles that will eventually leave for the ocean. Others are resident, shoreline fish, such as blennies, and gobies.

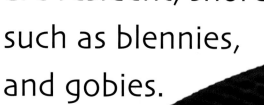

Rockpool blenny

Nora the Naturalist says:
Some types of tide pool fish, such as rockfish, have sharp spines in their dorsal fin. In some species these can be poisonous.

Most tide pool fish feed on small crustaceans, although some larger species eat other fish as well.

The types of tide pool fish vary, depending on the ocean they live in.

Most tide pool fish are between 4–12 inches (10–30 cm) in length.

These fish are members of the ray-finned fishes. This means their fins are webs of skin with bony spines.

There are many tide pool fish, some of which have strange names such as butterfish, lumpsucker, and worm pipefish!

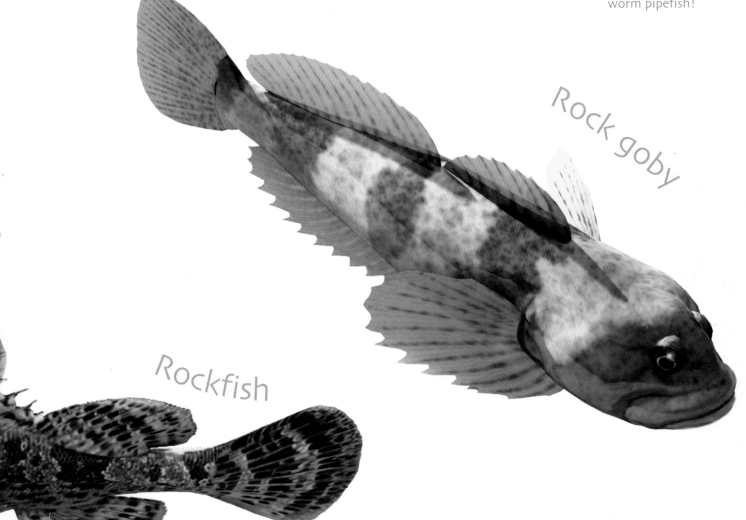

Rock goby

Rockfish

13

Starfish

Starfish are among the most well-known sea animals found on the seabed. They usually have five arms, although some species have many more. There are tubed feet that run along the underside of the arms.

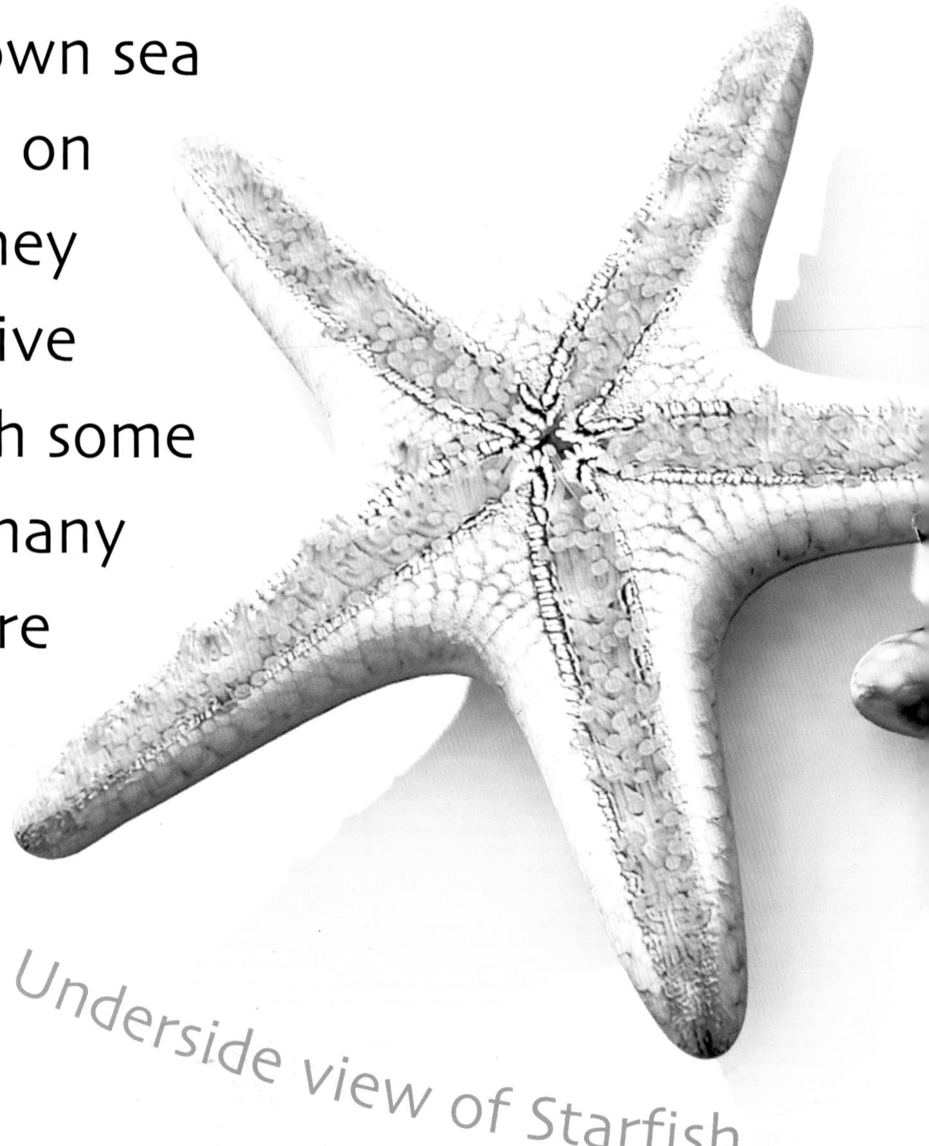

Underside view of Starfish

14

Starfish are found in all oceans around the world.

Starfish feed on animals living on the sea floor, such as clams, oysters, small fish, and snails. They may also eat **algae** and decaying plant matter.

Starfish grow to 4.5–9.5 inches (11.4–24.1 cm) in length.

Starfish belong to the family called **echinoderms**, which includes sea urchins, sand dollars, and sea cucumbers.

Starfish have a mouth on their underside, in the center. Some species can live up to 34 years.

Top view of Starfish

Nora the Naturalist says: There are around 1,600 different types of starfish. Some have the ability to regrow lost arms.

15

Jellyfish

Watch out for these sea creatures. They have dangerous stinging tentacles. They are often found washed up on beaches and in tide pools after storms at sea.

Black sea nettle

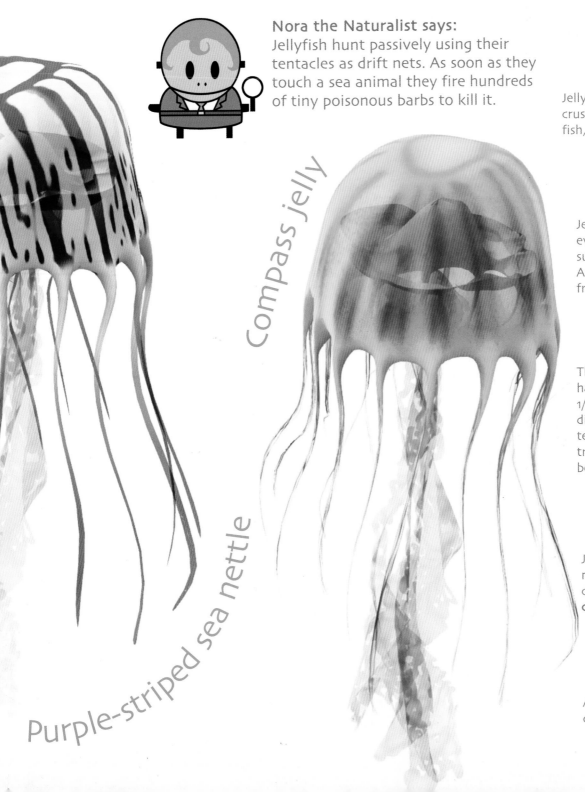

Nora the Naturalist says:
Jellyfish hunt passively using their tentacles as drift nets. As soon as they touch a sea animal they fire hundreds of tiny poisonous barbs to kill it.

Compass jelly

Purple-striped sea nettle

Jellyfish feed on plankton, crustaceans, fish eggs, small fish, and other jellyfish.

Jellyfish are found in every ocean, from the surface to the deep sea. A few jellyfish live in freshwater.

The smallest jellyfish have bell disks less than 1/16th inch (1.6 mm) in diameter, with short tentacles which they trawl around on the bottoms of rocky pools.

Jellyfish are the main non-polyp form of creatures of the **cnidaria** family.

A group of jellyfish is called a bloom.

17

Octopus eat small fish, shrimp, crab, lobsters, scallops, mussels, and clams.

Octopus are found in all the world's oceans.

The giant Pacific octopus can grow to an arm span of 14 feet (4.3 m). One of the smallest species is the California Lilliput octopus, which measures less than an inch (2.5 cm) across.

Octopus are members of the cephalopod family that includes squid, and cuttlefish.

Octopus are masters of disguise. They can change the color of their skin as well as its texture to look like crumbly rock, or spiky seaweed.

Octopus

Octopus are very intelligent sea creatures. They have eight arms and no bones, which allows them to squeeze through the smallest of gaps. If they are attacked they squirt out a thick, blackish ink in a large cloud.

Nora the Naturalist says:
An octopus can use a form of waterjet propulsion to escape from predators.

Octopus

19

Nora the Naturalist says:
The sharp spines can inflict a painful wound when they penetrate human skin.

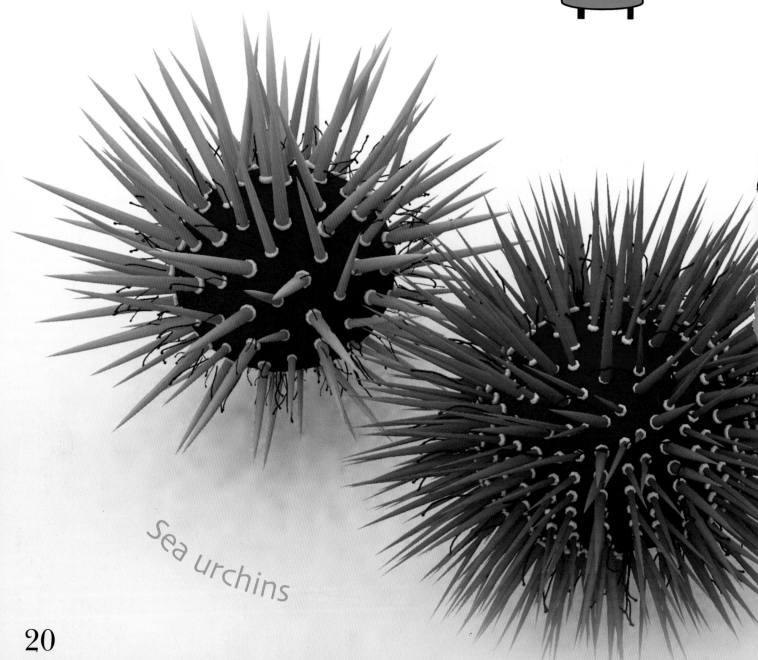

Sea urchins

20

Sea Urchins

Sea urchins are small, spiny, spherical animals. They have tube feet, similar to those of a starfish. Their long, sharp spines protect them from predators.

Sea urchin shell

Sea urchins feed mostly on algae, but can also feed on sea cucumbers, and other slow moving sea creatures.

Sea urchins are found in all the oceans around the world.

Sea urchins typically measure from 2.4–4.7 inches (6–12 cm) across. The largest can reach up to 14 inches (36 cm).

Sea urchins belong to the family called echinoderms, which includes starfish, sand dollars, and sea cucumbers.

Sea urchins' teeth are self-sharpening, and can chew through stone.

Most bivalves are filter feeders, capturing tiny food such as plankton from the water. Some species of sea snail feed on bivalves.

Nautilus are cephalopods with tentacles like a squid. They live in the deep oceans, and only their shells end up in the shallows.

These shelled sea creatures are members of the mollusk family.

Most of these mollusks grow to between the size of a finger nail to the size of a hand. Some, such as the giant clam, can measure as much as 47 inches (120 cm) across.

Sea mollusks are found in oceans all over the world.

Horn shells

Sea snails

Clam

Whelks

Sea Shells

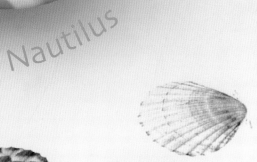
Nautilus

Many tide pools and beaches contain a variety of sea shells. They are the the remains of different types of sea creatures from sea snails and nautiluses to scallops and clams.

Nora the Naturalist says:
Curly shells, such as horn shells and whelks, are types of sea snail. Scallops and clams are animals enclosed by a shell in two hinged parts, called bivalves.

Scallops

23

Glossary

algae
Small, simple organisms.

cnidaria
Simple animals with stinging cells, such as jellyfish.

crustaceans
Mainly water animals with a segmented body, a hard outer skeleton, and paired, jointed limbs.

echinoderms
Sea creatures that include starfish, sea urchins, and sea cucumbers.

mollusks
The family of animals that includes squid, cuttlefish, octopus, and snails.

plankton
Microscopic organisms that live in water.

Index

crabs 4–5
crayfish 10–11

fish 12–13

jellyfish 16–17

lobsters 6–7

octopus 18–19

sea shells 22–23
sea urchins 20–21

shrimp 8–9
starfish 14–15

24